On the Way
to
The Man in the Moon.

Intercept Theorem to derive

maximum visible distance for planetary objects

based on Occultation of Sun and Moon.

Andrew Robert Chapman 21.8.2024

Edition: 8.9.2024

Prologue

The paper "Twinkle, Twinkle Little Star" ended with a mathematical paradox concerning the size of the Sun and Moon.

This (unplanned[1]) intermediate paper discusses that paradox in detail. It applies simple mathematics to calculate the maximum physical distance which heavenly objects can be seen, derived from their size and the (triangulation) ratio given by the Sun and Moon when viewed from Earth.

Book cover

A view of the Aug. 21, 2017, total solar eclipse from Madras, Oregon.

NASA/Gopalswamy[2]

---

[1] Planned was "The Man in the Moon", which will discuss how the distance of the Moon from the Earth's surface can be approximately estimated using light diffraction and the resolution of our eyes (or optical devices)

[2] https://science.nasa.gov/eclipses/media-resources/

Table of Contents

TWITTER Summary..1

Management Summary...2

Occultation: What are we measuring?....................................4

The paradox: Size does matter?..5

Trigonometry...6

Basic Proportionality/Intercept Theorem................................8

Diagrams: N.T.S. (Not To Scale)..10

Vanishing Point (max. visible distance).................................13

Mars Occultation..14

Conclusion...18

What Do I Know?...19

Dead or alive...21

Illustrations...22

Epilogue..26

Internet Hyperlinks...27

TWITTER Summary

From our viewing perspective on the surface of the Earth, the Sun is almost perfectly eclipsed by our Moon, meaning they have the same (relative) size.

The Moon has a radius of around 1,700 kilometres and orbits Earth at an average distance of 380,000 kilometres. The Sun has a radius of approximately 700,000 kilometres and Earth orbits it at a mean distance of almost 150,000,000[3] kilometres.

Knowing the sizes and orbits of the Sun and Earth/Moon we can form triangles (see Figure 1) and use trigonometry and/or intercept theorem to calculate the vanishing point[4], the distance from the Sun where both Sun and Moon will have a (relative) physical size of zero.

Surprisingly this vanishing point is not as large as one might imagine.

---

[3] This distance is also known as 1AU (Astronomical Unit)

[4] https://en.wikipedia.org/wiki/Vanishing_point

Management Summary

We are lucky to live in an era where mankind knows more about the world and universe than at any other time. We have access to machines and technology which, even 100 years ago, would have been the wild imaginings of fantasy and dreams.

Specifically relevant for this paper: Science has mapped the Solar System: The size of the Sun, its planets and moons and their orbits[5]. Mankind is busy mapping far away galaxies and star systems.

Famously six Apollo missions landed two astronauts on the Moon in the late '60s and early '70s and, more importantly, they'd dock the Lunar Module with the third astronaut in the awaiting and Moon-orbiting Command Module and return to Earth for splashdown.

So we, importantly for this paper, absolutely know *exactly* how far away and how big the Moon is.

---

[5] https://en.wikipedia.org/wiki/Solar_System#Orbits

The orbits invariably have varying degrees of eccentricity the limits of which are termed aphelion and perihelion. This paper will use a mid-point value, unless otherwise specified.

We are living in an age where the theoretical and practical sciences have calculated the speed of light, gravity, and sub-atomic bits of our Universe down to and beyond the Boson particle as well as discovering tiny, Earth-like and possibly habitable planets, in massive galaxies tens of thousands of light years away.

And mankind also has access to state of the art technology: powerful computers to help us with our scientific endeavours, electron microscopes, huge telescopes on earth and in low-Earth orbit, space rockets and stations and tens of thousands of expert scientists, technicians, programmers, engineers, professors and astronauts and astronomers who prepare, process and analyse the data for us. Of course mankind also has access to all the knowledge passed down through history; mathematicians, including Pythagoras' and/or Thales theorem, from 2,500 years ago.

Surprisingly the application of 2,500 year old mathematical formulae with modern day planetary data clashes irrevocably with our understanding of the Solar System and Universe.

The vanishing point of the Sun (and Moon) is such that it does not allow for the Sun's light to travel as far as the outer planets (not even to Mars) to illuminate them so we can see them, unaided, in our night sky.

But see them we do ….

Occultation: What are we measuring?

An occultation[6], an eclipse, is when a heavenly body moves behind another and is obscured from view. Most famously on Earth we see the Moon eclipsing the Sun but the Moon can also eclipse planets (and of course planets can eclipse planets etc. etc.).

We shall initially be concentrating on the eclipse of the Sun and Moon, which can be demonstrated using a practical experiment: The reader should grip a coin between thumb and forefinger and, standing about arm's length from the surface of a mirror, hold the coin in front of the face and move it to a position (closer to the face than the mirror) where the coin completely hides the reflected face and head.

If you find yourself focusing on solely on your reflection, or the coin, then close one eye and repeat the experiment. As soon as the coin covers your reflected face/head open your second eye.

In this paper your head represents you on Earth, the coin is the Moon and your reflected head is the Sun.

The coin is obviously smaller than your reflected head but its closeness to your eyes has the effect of it being the same **relative** size.

---
[6] https://en.wikipedia.org/wiki/Occultation

The below section, copied from "Twinkle, Twinkle Little Star", is the paradox which led to this paper.

The paradox: Size does matter?

The size of a light source makes no difference to the distance it travels. Assuming a constant medium[7] through which it travels, the brightness or luminosity is the sole deciding factor of how far beams of light travel.

Nevertheless, and as a separate exercise, we can calculate the point where the (relative) size of the Sun is effectively zero. We can do this by using high-school mathematics: Trigonometry and/or the Basic Proportionality/Intercept Theorem.

First consider Figure 1 illustrating our Sun (red) and Moon (grey):
AB: Sun's radius is (700,000 km).
MN: Moon's radius (1,700 km).
BN: The distance between Moon and Sun (150,000,000 km)[8].
ABN and MNB are both right-angles (90 degrees).

---

[7] Obviously a patch of fog or dust in the Earth's atmosphere will reduce the distance light travels.

[8] The distance between the Moon and Earth (380,000 km) can be ignored. The Earths orbit around the Sun, varying between 152.097,597 km and 147,098,450 km (a variation of 4,999,147 km) is a factor 13 larger than the distance of 380,000 km).

Trigonometry

Using Pythagorus' theorem (SOHCAHTOA) and the cosine rule and the three "known" sizes (AB, MN and BN) we can use the triangles (MBN and ABM) to calculate the length BC. The point C being that at which the (relative) size of the Sun is zero.

Effectively we are calculating the length BC but the length NC will suffice, as we can add it to (the known) length BN.

To calculate NC we need to know the angle NMC or MCN.

The angle NMC is exactly equal to the angle BAM.

To calculate the angle BAM we have to know the properties of the triangle BAM and for that we have to start with the calculation of the hypotenuse, using Pythagorus' theorem, of the right angled triangle BMN.

$BM^2 = MN^2 + BN^2$
$BM^2 = 1,700^2 + 150,000,000^2$
$BM^2 = 2,890,000 + 22,500,000,000,000,000$
$BM = 150,000,000.00963$ [9]

The (tangent) angle MBN = O (MN) / A (BN)
Tan MBN = 1,700 / 150,000,000
Tan MBN = 0.000011333
Angle MBN = 0.000649 degrees

---

[9] Because the Moon is so miniscule, compared to the size of the Sun, the length of the tangent (BM) is effectively the same as the distance between Earth and the Sun).

Using the cosine rule for triangle ABM (to find length AM).

Angle ABM = 90 − 6.49 = 89.99 degrees.
$AM^2 = AB^2 + BM^2 - 2 \times AB \times BM \cos(ABM)$
$AM^2 = 700,000^2 + 150,000,000.00963^2 -$
$2 \times 700,000 \times 150,000,000.00963 \cos(89.99)$
$AM^2 = 490,000,000,000 + 22,500,000,002,890,000 -$
$210,000,000,013,487 \times 0.000011333$
$AM^2 = 22,500,487,622,890,000$
$AM = 150,001,625$

Using the cosine rule for triangle ABM (to find angle BAM)

$BM^2 = AB^2 + AM^2 - 2 \times AB \times AM \cos(BAM)$
$22,500,000,002,890,000 = 490,000,000,000 +$
$22,500,487,622,890,000 - 2 \times 700,000 \times 150,001,625 \times \cos(BAM)$
$-977,620,000,000 = -210,002,275,561,158 \times \cos(BAM)$
$0.4655 = \cos(BAM)$
Angle BAM = 89.7332 degrees
Angle AMB = 0.2673 degrees (180 − 89.7332 − 89.99)

Angle BAM == Angle NMC

Length NC can be calculated using tangents.
Tan NMC = O (NC) / A (MN)
Tan 89.7332 × 1,700 = NC
NC = 365,173 kilometres

Basic Proportionality/Intercept Theorem

This mathematical theorem is much simpler than trigonometry. The theorem effectively allows lengths of similar triangles to be expressed (and solved[10]) as an equation.

AB/MN = BC/NC (where BC=BN+NC)

NC = MN.BN/AB-MN

NC = 1,700 x 150,000,000 / 700,000 – 1,700
NC = 365,173 kilometres

Both the trigonometry and Intercept Theorem calculations are included as a LibreOffice CALC[11] spreadsheet (Trigonometry.ods) in the cloud (see Epilogue).

The result of the calculation is unexpected because the vanishing point, C, of both Sun and Moon (365,173 km) should be (much, much) greater than the Moon's Apogee (405.400 km) but it is only slightly larger than the Moon's Perigee (362,600 km).

---

[10] See Figure 2

[11] Beware! The spreadsheet uses the German/European decimal comma notation.

Tweaking the results

The LibreOffice CALC spreadsheet is also an excellent way to "tweak" the sizes in order to see the effect on the vanishing point.

As mankind has flown to, orbited around and landed on the Moon multiple times, then the discrepancy must be with the Sun's data.

Reducing the size of the Sun and/or increasing the distance between the Sun and the Earth increases the vanishing point.

But tweaking (whether Sun or Moon data) implies that we have a fundamental problem with the data.

Diagrams: N.T.S. (Not To Scale)

The sizes and distances of the Solar system planets are literally astronomical and the human mind cannot easily comprehend them. It helps to scale the sizes and distances down to something we can literally measure ourselves. In this manner we can better envisage the relative sizes and distances.

We will be comparing a Sun (scaled to 140 cm) to Earth (scaled to 1.4 cm) and the Moon (scaled to 0.35 cm). Comparing an articulated lorry tyre (Sun) to a tomato seed (Moon).

Diagrams (see Figure 3) representing the Sun, planets and universe are NEVER, EVER even remotely to scale. (Technical) drawings are annotated with N.T.S. (Not To Scale) to denote this.

As a practical exercise to demonstrate the futility of representing planetary bodies to scale for the reader, we shall try to actually draw a part of our Solar system on paper. We will use a scale of 1 mm representing the equivalent of 1000 km.

Take a sheet of A3 paper or, if you have no A3, two sheets of A4 placed end to end will suffice.

We're going to draw the Earth and Moon (to scale).

Draw a 1.3 cm (13 mm) line in one corner of the paper. This represents the diameter of the Earth (13,000 km). If you have a compass draw a circle, otherwise sketch a freehand circle or simply leave the line (-), plus (+) or asterisk (*) to represent the Earth's approximate circumference.

Our Moon (3,500 km) will be a 0.35 cm (3.5 mm) line/circle but where must it be drawn? Well 380,000 km (the Moon's orbit around the Earth), scales down to 38 cm. That is then on the second sheet of A4 paper or near to the opposite side of an A3 sheet.

If we were to draw the Sun using this scale it would be a disc with a diameter of 140 cm (1.4 metres) and it would be 150 (one hundred and fifty) **metres** away from our 1.3 cm Earth.

At this scale 150 metres represents one Astronomical Unit (AU) and, for those interested in sharing a pizza on Mars with Elon Musk in 2025[12] you would find Mars a further 75 metres from the Earth, 225 metres from the Sun. Mars has a 0.6 cm (60 mm) diameter.

---

[12] People will go to Mars by 2025. "If things go according to plan, we should be able to launch people probably in 2024 with arrival in 2025," said Musk.

As a reminder: to date, in the late '60s and early '70s, we've managed to travel a distance of 38 cm to the Moon from the Earth. Next up (next year, at the time of writing) mankind will travel nearly 200 times that distance, 7,500 cm, to Mars.

If we wanted to draw Pluto on this scale model we'd have to walk (run or maybe drive or cycle) a distance of 6,000 metres (6 km) to plot a 2 mm wide "circle" representing our solar system's outermost dwarf-planet.

Therefore all diagrams and illustrations of planets and stars which you see in books (and even on video) cannot ever be anything other than "N.T.S.".

Bear in mind that after just "150 metres" (1 AU) our Sun (1400 mm) is effectively as big as our moon (3.5 mm). Common sense tells us that, at this rate of reduction, the Sun cannot have a relative size for very much more distance, as it is already, at Earth, very close to zero.

Vanishing Point (max. visible distance)

A vanishing point is an artistic concept. It is a point on a two-dimensional medium at which all objects "vanish" (they effectively become a single point on the artist's canvas).

It certainly stands to reason that this concept also applies to our 3D Universe, as an object simply cannot remain visible for effectively, infinity[13]. Were this the case then all objects in the Universe would be visible, interfere with each other and we would see everything at once (effectively nothing).

Objects in our reality are "seen" by our eyes in the visible light spectrum[14], and light diffracts (decays) exponentially according to the inverse square law[15]. Our eyes also have a limited resolution restricted by an optimum number (and type) of optical nerves which can be physically placed and serviced behind our eyeballs[16] and the physical distance from an object reduces its size and detail.

---

[13] An "AI" search leaves no options off the table (see Figure 4).

[14] CGI artist Impressions of far away Galaxies are rendered from radio and microwaves received in radio-magnetic telescopes.

[15] Refer to "Twinkle, Twinkle Little Star"

[16] Discussed in depth in the upcoming paper "The Man in the Moon"

Mars Occultation

But the discussion of whether objects have a vanishing point or not is rendered moot because the planets we observe in the Earth's night sky have form and colours. With just a basic hobby telescope (80 mm Aperture and 500 mm Focal Length) Saturn and its rings can be clearly observed, although the vanishing point of the planet[17] has long been exceeded. At its closest to Earth it is an immense 1.2 billion kilometres distant. The (weak) reflected sunlight has to travel the distance back to Earth so we can see it!

In other words the planets have a very definite, even if small, relative size, meaning they obviously haven't reached their vanishing points or maximum visible distance.

Astrobiscuits[18] excellent YouTube video demonstrating the power of the £10,000 C14 HD Edge telescope (355 mm Aperture and 3910 mm Focal Length), captures a Mars occultation (Mars is eclipsed by the Moon).

Using JPG editing software on video stills it is a simple matter to cut the disc (which is Mars) and paste it on top of the image of the Moon to gauge how "relatively large" Mars is.

---

[17] Almost 13 million kilometres using the intercept theorem.

[18] https://youtu.be/GQEYGAJIzAM?t=432

I pasted 25 images of Mars (see Figure 5). But the captured video image of the moon doesn't represent a full quarter, so we can approximate that we'd maybe need 40 for a full quarter. Let's be generous and round up to 50.

The Moon has a radius of 1,700 kilometres which means that each of the (50) Mars images has a relative radius of about 34 kilometres (1700 km/ 50 km).

Substituting Mars' relative size along with its other properties into the Intercept Theorem table:

Basic Proportionality Theorem	
	Unit
AB	3.400 km
MN	34 km
BN	75.000.000 km
NC	757.576 km

The result (NC) is very close to the vanishing points in the Libre Office spreadsheet, calculated using the Sun/Moon ratios, where Mars is calculated to have a vanishing point of 730,345 kilometres.

But, although the results appear confirmatory, the above calculation uses dimensions which don't match the angles of the Sun/Moon ratios.

In other words, substituting data individually into the Intercept Theorem is deceptive. Because the Moon almost perfectly occludes the Sun, when viewed from Earth, the Sun has the same (relative) size as the Moon. The perspective calculations of all heavenly bodies must be based exclusively upon the angles of the triangles ACB and/or MCN in Figure 1).

The results shown in the table below under the title "Max. visible distance" exactly reflect this intercept theorem calculation:

	NC=BCxMN/AB		Distance from	
			Sun	Earth
	Max. visible distance	Planet radius	AU (nearest)	Kilometres
Mercury	537.018 km	2.500 km	0,30	105.000.000
Venus	1.288.844 km	6.000 km	0,70	45.000.000
Earth	1.353.287 km	6.300 km		
Mars	730.345 km	3.400 km	1,50	75.000.000
Jupiter	15.036.517 km	70.000 km	5,00	600.000.000
Saturn	12.458.829 km	58.000 km	9,00	1.200.000.000
Uranus	5.370.185 km	25.000 km	18,00	2.550.000.000
Neptune	5.284.262 km	24.600 km	30,00	4.350.000.000
Pluto	255.191 km	1.188 km	30,00	4.350.000.000

The other columns reflect WIKIPEDIA planet sizes and distances from the Sun (AU) or Earth (kilometres). The discrepancies are colossal but let us study Mars in detail.

Mars, according to the results of applying the intercept theorem in the above table, has a vanishing point (maximum visible distance) of 730,345 kilometres. It cannot be seen after that distance. Period. Never-mind its official distance of 75,000,000 kilometres (one-half of an A.U. between it and Earth), because it has zero (relative) size.

The fact that Mars can not only be seen in Earth's night sky, but, compared to our Moon, can be estimated to have a relative radius of about 34 kilometres, means that it (obviously) must be closer (to Earth) than its vanishing point.

If we substitute the intercept theorem's vanishing point length (730,345 km), the radius of Mars (3,400 km) and its approximate relative size we observe from Earth (34 km) into the intercept theorem:

AB/MN = BC/NC
3400/34 = 730,345/NC
100NC=730,345
NC=7,303.45 kilometres

We can thus calculate that Mars is about 723,042 (730,345 - 7,303) kilometres from Earth. This implies that when the observer moves a mere 7,303 kilometres away from Earth, then Mars can no longer be visible.

The massive orbital variations of both Mars[19] and Earth[20] easily exceed 7,303 km by a factor of 700 to 6,000! Which means we would only be able to see Mars when Earth is "close" to it: The perihelion of Earth and the aphelion of Mars.

---

[19] Difference between ap- and peri- helion is 43,000,000 kilometres.

[20] Difference between ap- and peri- helion is 5,000,000 kilometres

This is obviously not the case and the relative size of Mars remains constant[21], despite fluctuations of millions of kilometres between Earth and Mars as they orbit elliptically around the Sun.

In fact none of the Solar system's planets suffers from "visibility"[22] problems due to the variations of the orbits.

Conclusion

The calculations have been checked, rechecked and double-checked. The conclusion of this paper is that the size and distance data for the Sun, Moon and planets does not apply itself well to the results of the intercept theorem.

Ideally a method of confirming the distance to the Moon is required by the layman and the paper "The Man in the Moon" will attempt to do just that.

---

[21] Or at least Mars always remains visible in Earth's night sky.

[22] Certainly not the 9 AUs out to Saturn

What Do I Know?

I obviously know what I was taught in school, including trigonometry. After school I perused a couple of books on astronomy and watched the odd documentary. As astronomy is not a subject of great personal interest to me I'd posit I know about as much as the average layman.

Nevertheless, taking a step back and looking at the world through a "child's eyes" can sometimes help remove the "blinkers" from the eyes which blind the specialists who are so busy working on at the cutting-edge of their science that they pay no more heed to the simple, "settled" stuff.

But my question, at the start of this section was "What do I know?" not "What have I learned and been told?"

I "know" what I can experience and prove to myself.

I "know" as much as you and, effectively, all the world's population. The same as all but a handful of (now) old (or dead) men who flew to the Moon.

I can confirm that the Sun is the same relative size as the Moon

That's really about it!

That's all I can claim that I know.

I have no idea how big those discs in the sky really are. How far away, or how hot or luminous. I have no idea if our Sun really is a massive, fiery, flame-spewing fireball 150,000,000 km away Those are things I "know" only because I have been told or taught them. The Sun could just as easily be a (much) smaller, not so fiery and hot and much closer source of light and heat.

The same applies to the Moon and the planets.

The planets are eclipsed by the Moon, therefore the Moon is closer to the Earth.

The simple mathematical intercept theorem requires just three knowns but, as we have read in this paper, those which are available in WIKIPEDIA do not yield results which marry and confirm the data of the Solar System planets.

Dead or alive

I offer a monetary reward to any person who can disprove this thesis. The reward of €250 will stand until 25.2.2025 and will be awarded to the first person to write to me with an explanation to this e-mail address:

ARCinfo33@proton.me

Should my thesis be disproved[23] I will extend and publish this paper with the "winning" counter-argument.

I will not speculate about a solution for the simple reason I don't have one.

---

[23] Disprove stipulates using mainstream scientific and mathematical theories and data alone. Ideally those which can be found in WIKIPEDIA.

Illustrations

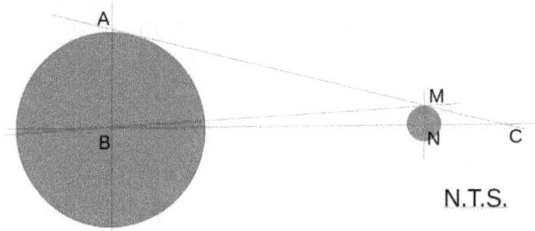

N.T.S.

Figure 1 Sun (AB=700,000 km), Moon (MN=1,700 km), Earth's orbit (BN=150,000,000 km), vanishing point C. Maximum visible distance (BC[24]).

Heavenly bodies arranged in size (radius) order			
	Radius (km)	MVD (km)	mDfE (km)
Pluto	1.100	255.191	4.350.000.000
Moon	1.700	365.173	380.000
Mercury	2.500	537.018	105.000.000
Mars	3.400	730.345	75.000.000
Venus	6.000	1.288.844	45.000.000
Earth	6.300	1.353.287	
Neptune	24.600	5.284.262	4.350.000.000
Uranus	25.000	5.370.185	2.550.000.000
Saturn	58.000	12.458.829	1.200.000.000
Jupiter	70.000	15.036.517	600.000.000
Sun	700.000	150.365.173	150,000,000
MVD	(Maximum Visible Distance) as calculated by the intercept theorem		
mDfE	(minimum) distance from Earth		

Table 1. Common sense implies that, if the Sun is already as relatively small as our Moon after 150,000,000 km, then Jupiter, with a tenth of the Sun's radius, will have a MVD a tenth that of the Sun's. The above table confirms this expectation (15,036,517 km) whereas WIKIPEDIA states that the planet is 40 times this distance from the Earth.

---

[24] Strictly speaking BC – AB, as we see the outer surface of the Sun.

$$\frac{AB}{MN} = \frac{BC}{NC} = \frac{(BN+NC)}{NC}$$

Cross multiply

$$AB \cdot NC = MN \cdot (BN+NC)$$

Solve brackets

$$AB \cdot NC = MN \cdot BN + MN \cdot NC$$

Divide by NC

$$AB = \frac{MN \cdot BN}{NC} + MN$$

Subtract MN

$$AB - MN = \frac{MN \cdot BN}{NC}$$

Multiply by NC and divide by AB-MN

$$NC = \frac{MN \cdot BN}{AB - MN}$$

Figure 2 Solving the Basic Proportionality/Intercept Theorem equation for NC

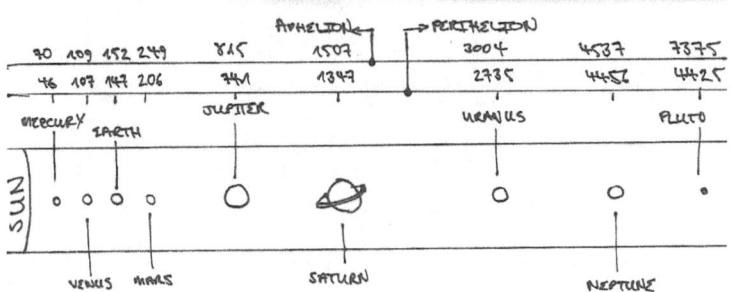

Figure 3 Trust me. This is definitely NTS

What is the maximum visible distance

Based on the search results, the maximum visible distance varies depending on the conditions:

- **Optimal conditions**: In ideal circumstances, a person with normal vision can see up to 20 kilometers (12 miles) on a clear day, according to one source.
- **Unobstructed view**: On a clear night, it's possible to see the Andromeda galaxy, which is approximately 2.25 million light-years away. However, during the day, dust, water vapor, and pollution in the air typically limit visibility to 20 kilometers (12 miles) or less.
- **Tall vantage point**: From a high altitude, such as the top of Mount Everest, the maximum line of sight can theoretically reach 339 kilometers (211 miles), but in practice, cloud cover often obstructs the view.
- **Specific location**: The Calgary Vision Centre mentions a road with a straight and flat terrain, allowing for a maximum visible distance of approximately 443 kilometers (275 miles) under optimal conditions.
- **Theoretical infinity**: Some sources suggest that, in theory, the human eye can see infinitely far, as there is no physical limit to the range of vision. However, this is subject to the presence of obstructions and atmospheric conditions.

In summary, the maximum visible distance varies from approximately **12 miles (20 kilometers) in optimal conditions to potentially infinite distances under ideal circumstances**, with various factors influencing the actual range of vision.

Figure 4 An "AI" summary of maximum visible distance? From 20 km to infinity. That certainly narrows it down …

Figure 5 Twenty-five Cut-and-Pasted Mars images along the bottom of the video still. As this isn't the full radius of the Moon the paper "guesstimates" that 50 would be required.

Epilogue

      My research paper "Twinkle, Twinkle Little Star" not only left questions unanswered for the distance light emitted from stars can travel its final section "The paradox: Size does matter?", the basis for this paper, also calls into question the size and distance of Solar system objects.
      With so many data being called into question in this and the preceding paper, a follow up dissertation will attempt to provide answers by calculating the distance between the Moon and the Earth, without having to resort to the impracticability of building a personal space-rocket.

A PDF of the latest version of this book/paper is located at:

https://www.mediafire.com/folder/xvuu4g73xr07o

      There you will also find other relevant EXCEL files and also TXT/RTF answers which I have received and my response to them. Thus keeping this publication focused and minimal.

Internet Hyperlinks and credits

All links in this paper can be found in the below QR code and were active at the time of writing and initially going to press.

Credits

I wish to express my explicit thanks to:
Anthony George Chapman: proof-reading, corrections, suggestions and discussions concerning the content of this paper.
Peter Wätjen: The Intercept theorem solution.

Other books by the Author

Please visit AMAZON to discover other books.

Science Paper Series
ISBN-13: 979-8872921790: Twinkle, Twinkle, Little Star
z/OS Guide Series
ISBN-13: 979-8601937375:
COBOL: Optimised & Maintainable Application Programming
ISBN-13: 979-8602445213: JCL – STEP by STEP
ISBN-13: 979-8882582233: Practical z/OS HLASM

Songwriting Series
ISBN-13: 978-0463202951: F.A.T.E: ALPHAs & BETAs
ISBN-13: 978-0463492048: PackEis: Lyrics and BETAs
ISBN-13: 978-0463844366: WildßcreW: A Graphical Revelation
ISBN-13: 979-8354626465: G.A.S.C. and I

Lyrics and Poetry Series
ISBN-13: 978-0463277454: 2011 & earlier: Lyrics & Poems
ISBN-13: 978-0463393758: 2012: Lyrics & Poems
ISBN-13: 978-0463205938: 2013: Lyrics & Poems
ISBN-13: 978-1370272846: 2014: Lyrics & Poems
ISBN-13: 978-0463610473: 2015: Lyrics & Poems
ISBN-13: 978-0463923771: 2016: Lyrics & Poems
ISBN-13: 978-0463472866: 2017: Lyrics & Poems
ISBN-13: 978-0463173268: 2018: Lyrics & Poems
ISBN-13: 978-0463388174: 2019: Lyrics & Poems
ISBN-13: 979-8582846918: 2020: Lyrics & Poems
ISBN-13: 979-8832205168: 2021: Lyrics & Poems
ISBN-13: 979-8832205168: 2022: Lyrics & Poems

Corona Pandemic Series
ISBN-13: 979-8859838721: Saving Granny

www.ingramcontent.com/pod-product-compliance
Lightning Source LLC
Chambersburg PA
CBHW070958220526
45471CB00007B/3088